WOOL

HARVEST TO HOME

Lynn M. Stone

Rourke Publishing LLC
Vero Beach, Florida 32964

www.rourkepublishing.com

PHOTO CREDITS:
All photos © Lynn M. Stone except p. 13 courtesy Pendleton Woolen Mills

EDITORIAL SERVICES:
Pamela Schroeder

Library of Congress Cataloging-in-Publication Data

Stone, Lynn M.
    Wool / Lynn M. Stone.
        p. cm. — (Harvest to home)
    ISBN 1-58952-131-5
    1. Sheep—Juvenile literature. 2. Wool—Juvenile literature. [1. Wool. 2. Sheep.] I. Title

SF375.2 .S77 2001
636.3'145—dc21   2001031672

Printed in the USA

TABLE OF CONTENTS

WOOL

Wool is the thick, soft hair that covers some animals, like sheep. This book is about sheep wool. Sheep wool is used to make warm and beautiful clothing and blankets.

Different **breeds** of sheep grow different types of wool. For example, the **fibers**, or threads, of wool may be just 1/2 inch (about 1 centimeter) long on one breed. Another breed may grow wool fibers 14 inches (36 centimeters) long!

Sheep grow a warm, dense coat of hair called wool.

Wool fibers may be flat or wavy. Some have up to 30 waves per inch (12 waves per centimeter). Wool has different natural colors. Sheep may be white, brown, black, or other colors.

Wool is valuable for more than just its good looks. First, it is warm and lightweight. It does not wrinkle easily, and it keeps off light rain.

Wool can be **dyed** in bright colors and not fade. Wool will burn, but not easily. It can also be reused.

These homemade wool items have been dyed different colors.

SHEEP

The **domestic** sheep of today have wild **ancestors**. Domestic sheep have been around for a long time. Farmers began taking sheep from the wild as long as 8,000 years ago! The first domestic sheep to reach North America were brought from Spain around 1519.

Some breeds of sheep have horns. Others do not. Both **rams** and **ewes** grow horns in some breeds.

Merino ewes are hornless but the rams shown here have thick, curled horns.

Ewes are mother sheep. They weigh from 100 to 225 pounds (45 to 102 kilograms). Rams, the fathers, weigh 150 to 350 pounds (68 to 159 kilograms).

A ewe can give birth when she is about 1 1/2 years old. She may have one, two, or even three lambs at once.

A mixed-breed ewe stands with her many-colored twins.

Yarn is made by spinning—drawing out and twisting strands of wool.

A weaving machine at a wool factory makes yarn into cloth.

Some breeds of sheep are raised for meat. Other breeds, like the curly-coated Merino, are raised for wool. The Corriedale is a breed raised for both meat and wool.

Wild sheep live in some of the highest, rockiest country on Earth. It's no wonder that domestic sheep do, too. Sheep raised on some of the cold, rocky islands off Scotland live partly on a diet of seaweed. In the United States, sheep in the West live on short grasses in dry mountains.

Icelandic sheep graze on the seaside cliffs of Iceland. Their long, colored fleece is world famous.

SHEEP FARMS

Most sheep in the United States are raised west of the Mississippi River. Texas raises more sheep than any other state. California, Wyoming, Colorado, and South Dakota follow Texas.

Many sheep in the West live in groups of several hundred on broad, open range land. Other sheep are raised on farms. Sheep farms may have a small flock or several hundred sheep.

This flock of sheep is being raised for its wool on a Wisconsin farm.

MAKING WOOL

Farmers get wool from their sheep by **shearing** it off. Shearing a sheep is like giving it a "haircut." Wool grows back, just as your hair does.

Wool is sheared with an electric clipper. The person who shears tries to take off the wool in one, blanket-like piece called a **fleece**.

A Vermont farmer shears a muddy young ewe.

The making of wool clothing begins with washing and cleaning the fleece. The natural **lanolin** wax on the wool is usually taken off. Lanolin is used for beauty aids and lotions.

After being cleaned, the wool is dyed. Dyeing adds color.

The wool is then combed into thin sheets or webs. From these sheets the wool is separated into long, thin "ropes" called strands.

A home spinner combs wool into a fine sheet, or web, in a process called carding.

The strands of wool are combed and spun. During the spinning process, wool is twisted into finer strands called yarn. Two sets of yarns are then woven together on a **loom**. Weaving the yarn makes cloth.

The final steps are called "finishing." Finishing uses heat, water, and force. Together these steps prepare the wool cloth for tailors. Tailors make the cloth into clothing.

GLOSSARY

ancestor (AN ses ter) — those of the same family who came before; early relatives

breed (BREED) — within a certain group of domestic animals, one special type, such as *Merino* sheep

domestic (deh MES tik) — the kinds of animals that have been tamed and raised by people for thousands of years, such as sheep and cattle

dyed (DYD) — to have had color added

ewe (YOO) — a female sheep

fiber (FY ber) — a single strand or thread

fleece (FLEES) — the wool sheared from a sheep

lanolin (LAN eh lin) — a natural wax on sheep wool

loom (LOOM) — a machine that weaves wool

ram (RAM) — a male sheep

shearing (SHEER ing) — the taking of wool from a sheep

INDEX

Further Reading

Kalman, Bobbie D. *Hooray for Sheep Farming!* Crabtree, 1997

Stone, Lynn M. *Sheep*. Rourke Publishing, 1990

Websites To Visit

www.sheepusa.org

www.tsgra.com (Texas Sheep and Goat Raisers Association)

About The Author

Lynn Stone is the author of more than 400 children's books. He is a talented natural history photographer as well. Lynn, a former teacher, travels worldwide to photograph wildlife in its natural habitat.